BEE HOTEL

The Author

Melanie von Orlow was born in Germany, and studied biochemistry and biology in Berlin. She now works as a freelance biologist in Berlin, capturing bee swarms and providing relocation and advisory services. As spokeswoman for Hymenoptera, a division of the German Nature and Biodiversity Conservation Union (NABU), she lobbies nationwide for bees and wasps, and conducts training courses, lectures and workshops for advisors and relocators. She maintains a small honey farm in north Berlin where she runs courses for the next generation of beekeepers.

First published in September 2019

British Library Cataloguing in Publication Data
A catalogue record for this book is available from the British Library.

ISBN 978 1 78521 658 9

Library of Congress catalog card no. 2019934679

Published by Haynes Publishing,
Sparkford, Yeovil, Somerset BA22 7JJ, UK
Tel: 01963 440635
Int. tel: +44 1963 440635
Website: www.haynes.com

Haynes North America Inc.
859 Lawrence Drive, Newbury Park,
California 91320, USA

Printed and bound in Malaysia

BEE HOTEL

ALL YOU NEED TO KNOW IN ONE CONCISE MANUAL

30 DIY INSECT HOME PROJECTS

Melanie von Orlow

Contents

Insect-friendly gardens

Did you know that if you have a garden or balcony, you're also a 'bee keeper'? Wild bees, solitary wasps, earwigs and green lacewings are unassuming, peaceful and inconspicuous neighbours.

WHO LIVES IN THAT SCREW HOLE?

You'll have to look carefully to notice bees whizzing into the flower box or window frame. Often, all that can be seen is the pollen they deposit in garden furniture screw holes. Solitary wasps are even more unobtrusive, usually going about their business in muted black and therefore mostly not identified as wasps in the first place. Just like wild bees, they come in all

⬇ Red mason bee on reeds carrying home yellow pollen on her belly.

sizes, small as an ant, large as a hornet. Among them are many valuable allies in a gardener's battle against aphids and caterpillars, which they target to provision their nests. These 'solitary' insects don't form colonies but do everything themselves, from building nests and laying eggs to provisioning.

The relatives of our social honeybees and 'cake wasps' (common wasps) provide vital services in the garden as pollinators and exterminators of caterpillars and aphids. Solitary bees and wasps are

also far more peaceable than their sociable relatives. Many species are either stingless or their sting is too weak to penetrate our skin. Even the few species that could inflict a painful sting refrain from doing so. With a lifespan of only four to eight weeks, their life is simply too short to risk anything when defending their nest. This makes these insects ideal 'pets' for all who enjoy watching nature in their own garden. Wild bees, etc. let parents provide their children with an exciting and safe experience of nature.

THREE WISHES . . .
Solitary bees and wasps build a huge variety of nests: some build resin nests on rocks; others dig tunnels in pavement gaps; yet others create elaborate mud structures on plants. All these species have one thing in common, however: they don't venture far. All they need has to be available within a radius of only a few hundred metres. Depending on the species, this means they need this close by:

- certain **forage plants** (box on page 11) for supplying their larvae with pollen (in the case of wild bees) or prey (in the case of female wasps);
- **building material**, mud, plants with long plant hairs, resin or plant oil, petals or leaves of certain plants from which they cut pieces;
- favourable **nesting sites**, wood with holes, suitable spots on the ground, mud or loam.

You can be 'fairy godmother' to many species by fulfilling all three wishes at once in only a few steps! With the bug hotels described in this book you can provide a home for ground dwelling species (about a

⬆ DIY is best – your nest will soon be fully booked.

third of all European wild bee species) and also attract other beneficial insects such as green lacewings, earwigs and butterflies.

Ideal homes for insects
A home for wild bees can be made anywhere, even in towns and cities. You can make bug hotels in many locations as long as a few basic conditions are right.

↑ Depending on the species, sweat bees are happy to live in various kinds of nests.

A general rule: the nest should be sunny, and it must be dry. Although shady nesting sites will still attract inhabitants, nesting places that are in damp spots (on the weather side, in treetops, near the ground) are unpopular. Also, insects appreciate firmly secured nests that don't sway. Ideally you should set them up on vertical surfaces such as exterior walls, balcony parapets, walls and tree trunks, or firmly secure nesting houses or large nesting walls in the ground.

Some species emerge late in the year; it makes sense to regularly add new designs to provide space for them too. Although not all species clean their tunnels, much preferring first occupancy, many species do clean their nests. So you may find that regular cleaning is unnecessary; and be sure not to remove any inhabited tunnels when adding and replacing elements.

NATURALLY COSY WOOD

Bug hotels made of wood are best built using well-seasoned solid timber, preferably hardwood. The most durable wood is robinia (false acacia), oak, and wood from fruit trees. Coniferous timber is less suitable since the resin ducts put the insects off and it's less weather-resistant. As a rule, wood for nests should be untreated and remain so: neither pressure-treated wood (identifiable by its green tinge) nor wood coated with oil or paint (even if environmentally friendly) are suitable. However, you can decorate the roof and frame of your hotel with environmentally friendly paints.

Be sure to use sharp tools for drilling and shaping to avoid any protruding wood fibres that could get in the way. Tap out the holes well after drilling. You can remove the last remaining fibres around the drilled holes using sandpaper (80–120 grit) or a round file.

The result will be smooth, clean nesting tunnels such as those left behind in nature by wood-dwelling beetle larvae for wild bees to use.

Maintenance and repair

Bug hotels should remain outdoors throughout the year. If a hotel that's been used for many years remains uninhabited for a long time in spite of undamaged tunnels in good condition, in the winter you should tap it out on a hard surface with the openings facing down to remove any dead larvae or remaining mud – it might then attract new inhabitants again the following spring.

Frequently seen, but not ideal, are holes drilled into end grain: the classic wooden cross-section with drilled holes dries out over the years, resulting in radial cracks. Tunnels featuring such cracks are not very popular with insect populations.

It's usually better to drill into rip-cut wood that has been stripped of bark. Always leave gaps of 1–2cm between holes. The drilled holes should be horizontal, closed at one end and about 10 times longer than their diameter (eg 8cm length to 8mm diameter). They can be arranged any way you like: scattered, or forming a pattern. Choose diameters ranging between 3 and 9mm – many hotels feature tunnels that are much too wide and never attract inhabitants. Many insects, especially wasp species preying on aphids, prefer small diameters between 3 and 5mm, so these should be provided in larger numbers. Provide tunnels of 5–6mm and larger and you can expect mason bees as regular guests. The nesting tunnels don't have to be round – square and semi-circular holes are equally acceptable.

WALK-IN PASSAGES MADE OF REEDS AND BAMBOO

Reeds or rushes are easiest to obtain in the form of reed screens, purchased from DIY shops and garden centres. These outlets also sell bamboo canes as growth supports.

Both the reeds and bamboo canes should be intact – that is, not squashed or split lengthwise. Follow these steps to obtain neatly cut edges:

1 Cut reeds to size using sharp garden shears (pruning shears).
2 Cut bamboo canes with a saw with a fine blade.
3 Cut both reeds and bamboo canes so that each piece ends with one of the thicker nodes occurring at regular intervals along the cane – this means it is naturally closed at one end, which is important for prospective inhabitants.
4 Alternatively, seal open-ended bamboo canes with cotton wool or some mud, or by affixing them on a wooden panel.

⬇ Wood with drilled tunnels offers countless design possibilities.

↑ Bamboo is readily colonised and makes a particularly durable material.

Diameters vary naturally along the length of canes and stalks. In the case of bamboo, it may be necessary to remove the pith with a wire; pith-filled canes are not suitable for nesting when arranged horizontally even though this is how they are frequently used in bug hotels.

PITH-FILLED CANES AND STALKS FOR DIY INSECTS

Pith-filled canes and stalks, when placed vertically, are good for species that like to chew into the pith to build their own nesting tunnels. They are used only once and should be renewed after the offspring has hatched (identifiable by the opened seal on the cut in the following year). If you're unsure whether the insects have already hatched, store the canes or stalks in a dry location in the garden.

You can use the stalks of numerous trees and shrubs such as common mullein (*Verbascum*), globe thistle (*Echinops*),

Protection from beaks

Birds are likely to clean out the canes and stalks if you just insert them into the box. So glue them to the back panel using waterproof wood glue and cover the front with a bird-proof grate that won't block the insects, for example a piece of fine chicken wire mesh.

Nectar and pollen sources

Provide a well-stocked garden for your bees: all Stachys species (eg lambs' ear), dead-nettles (*Lamium*), bellflower (*Campanula*), loosestrife (*Lysimachia*) and *Sedum* provide excellent bee forage. Other forage plants include all kinds of herbs (eg mint, allium, thyme, oregano, hyssop) and early flowering plants such as spring crocus and bluebells (*Scilla*). Native flowering plants are always a good choice.

↑ The ivy bee is a species of plasterer bee, they readily accept nesting bricks.

blackberry, raspberry, butterfly bush (*Buddleja*) and elder. Fit the stalks individually (not in a bundle) and vertically, for example along the garden fence using wire. They may vary in length. The important thing, of course, is access to the pith by way of at least one cut since most species are unable to access it themselves.

BUILT IN STONE – CLAY AND BRICKS

The holes of most perforated bricks are too big to be of interest to solitary bees and wasps. Better, but not always easy to obtain, are extruded interlocking tiles (see page 40 for alternatives). This type of beavertail roof tile features longitudinal tunnels that after quick deburring of the entry holes with a stone drill and sealing of tunnels at the back with clay provide a simple nesting site. These very long tiles may also be broken in half to provide space for even more insects.

Potter's clay with properly drilled tunnels (see page 8) also makes weatherproof nesting sites after firing; however, nesting tunnels in mineral materials seem to be less popular than those in organic material

such as wood, bamboo canes and reeds. Wood, concrete and breeze blocks can also be adapted for nesting sites.

No home for wild bees

Frequently used in ready-made bug hotels, but rarely colonised by wild bees and other insects: spruce and pine cones, horizontally stacked pith-filled stalks, wood shavings or straw behind wire mesh, perforated bricks and glued-on snail shells. Here only the odd spider will make its home.

INSIGHTS INTO THE NURSERY

Hotels with 'windows' for looking inside the nesting tunnels should be fully blacked out and opened only briefly for checking. Proven designs are grooves in wood with one side covered by acrylic glass. This allows moisture to evaporate to protect the eggs and larvae from fungal infection. Acrylic glass cracks easily and therefore requires careful sawing with a fine saw and fine pre-drilling. It's best to have the acrylic glass pane cut to size right away at the suppliers.

FOR BUILDERS: CLAY AND LOAM

Clay and loam (mud) walls are very attractive nesting sites as well as sources of building material for many species. The

⬇ Mud is an important building material for insects, but is often in short supply

species nesting in mud actively chew their tunnels into the material.

If you don't have natural loam or clay deposits in your area, you can work with fine clay skim finish or lean pottery clay from specialist retailers. It's important to make the clay 'lean' using very fine sand (*eg* paving sand): once dried, the result should be soft enough to easily scrape off with your fingers – then future inhabitants, too, will be able to work well with it. Find the right mix by trying out various small batches. These mixtures do not require reeds, stones, gravel or straw.

Pour the mud into clay or Eternit planters, hollow-core blocks or wooden troughs and compact it well. Then you can drill several shallow holes of only 1–2cm in length and about 6–8mm in diameter with ample space between the holes to provide the insects with a starting

↑ Sand bees are ground nesting insects, burrowing into the soil to build their hives.

point. Set up the filled troughs so that the mud wall is vertical.

GOOD SOIL FOR GROUND NESTING INSECTS

Most wild bee and solitary wasp species don't colonise ready-made tunnels but dig their own nests in the ground. While ground is available everywhere, it's often too densely vegetated, too shady, too damp or too soft to offer these insects a home. It's often easier to protect existing nesting colonies than to attract new bees – observant eyes will soon identify the typical sand heaps in pavement cracks as wild bee nests.

Loosely vegetated, dry, sandy but compacted soils in a sunny location are typical nesting sites: prior to digging, watering and planting you should check whether young wild bees might be growing up there. If you deposit some empty snail shells there you'll provide additional nesting opportunities for certain species. Incidentally, it's no use gluing snail shells

on to nesting walls: these bee species look for nests close to the ground and prefer to perfectly position the shells themselves.

DEAD WOOD

Provided it's in a sunny location, dead wood (starling nest boxes, wood from pruning fruit trees, logs) makes a natural nesting site for many wild bees. Vertical arrangements are highly desirable and simple to make: don't cut down dead fruit trees, only remove the branches and leave the trunk standing. After a few years you might even see violet carpenter bees here that bore deep into the rotting wood to create their nesting tunnels.

About this book

This book focuses largely on hotels for solitary bees and wasps; however, you can also provide a home for social bumblebees and hornets (more about this from page 69 onwards). It's also easy to attract aphid-eating ladybirds, green lacewings and earwigs as well as colourful butterflies. You'll find hotels for these insects from page 84 onwards.

Chapter 1
Hotels for solitary bugs

Solitary bees and wasps

'Busy as a bee' is a phrase that best applies to solitary bees and wasps, who wholly dedicate their short lives to the building, provisioning and sealing of breeding cells. They are easy to spot in the garden, in various makeshift temporary homes, such as screw holes and little spaces within masonry or wooden structures.

↑ This wild bee diligently walls up the entrance to her nesting tunnel.

In contrast to honeybees, solitary bees live as individuals and are classified as 'wild bees', along with bumblebees. They neither build huge nests nor produce honey, but they are efficient pollinators.

THE BEE MATING SEASON

The bees emerge between March and September, depending on the species. The males usually hatch a little earlier and await the females at popular forage plants or even directly at the nests. You might spot one flying around restlessly in front of the holes and, with a little luck, even observe them mating there. The females don't waste time looking for distant nesting sites; they often start building their own nests near where they hatched.

PAPERING, BUILDING WALLS

The female bees construct the breeding cells. Some species actively bore their tunnels into wood, mud or the ground; others use existing holes. Depending on the species, they then cut out small pieces of leaves, roll them up, and carefully insert them into the tunnel. Others first plaster the walls with mud before depositing pollen (often collected only from certain plants). Many species carry the pollen on their belly. You may be able to see how they enter the nesting tunnels with their abdomen first to cast off the pollen. The cell is sealed after egg deposition and work on the next cell starts. Inside the tunnels, they partition off individual cells by building walls.

Once the nest is complete, it gets sealed, and the seals are a sure sign that the tunnel is occupied. The whole process can take place within only a few days. The offspring then develops all on its own, emerging from the tunnel the following year.

Individual front doors

The nest seals of solitary bees differ from species to species. The insects create them using thin silk, leaves or mud, sometimes with inset stones, sometimes without. You'll gradually find out which door belongs to whom.

Solitary wasps

Europe is home to several hundred species of solitary wasps. They don't feed their offspring with pollen but with insects and spiders. Depending on the species, they will hunt caterpillars, aphids and other insects, which they paralyse with a sting.

Wasps can often be observed at the nests when they bring home their sometimes rather heavy prey; it's impressive how they then push these large insects into the tight tunnels. In other respects their lifecycle is the same as that of solitary bees.

Freeloaders and parasites: always lying in wait

You'll also observe numerous parasites at the nesting sites, for example brightly shimmering cuckoo wasps waiting for the right moment to smuggle their eggs into a bee's cell while it's still under construction. You might also see dark bee flies that in flight hurl their eggs into the

⬆ Got you! A solitary wasp with its prey.

open nesting tunnels. Mourning bees and blister beetles, too, might benefit from the industriousness of solitary bees and wasps at the nesting sites.

⬇ Cuckoo wasps often hover around nesting sites to foist their eggs on somebody else.

Hotel in a bucket

Time   2 hours
Build very easy

*Very easy to make, you
don't even need a drill.*

1 You can find small zinc buckets about 15cm high and 10cm in diameter sold as plant pots in the gardening section of almost every DIY shop. These buckets have the advantage of being weatherproof. Alternatively, you can choose a sheltered spot away from the rain and use a food tin which you can varnish in a colour of your choice.

2 Cut the bamboo canes or reeds as needed, following the instructions on page 9. Inserted vertically into the bucket, they should be no higher than about 1cm below the bucket's rim. Take care not to squash the reeds when cutting them.

3 Collect the bamboo canes or reeds into a bundle of the same diameter as the bottom of the bucket and tie it together with a rubber band. Squeeze a layer of aquarium silicone sealant on to the bottom of the bucket and firmly press the bamboo or reed bundle down into it. As the bucket is broader at the top, you'll now have to fill in the gaps: simply cut individual bamboo canes or reeds to the required length, apply a dollop of silicone at one end and stick into the bucket until evenly filled.

4 Allow the hotel to air for a few days to get rid of the silicone's vinegary smell as that could deter future inhabitants.

Either fix the bucket onto a wall with a screw through the bucket's base prior to filling or use a wire to hang it horizontally from a branch or fence.

Materials
- 1 or several small zinc buckets from a DIY shop or garden centre, height about 15cm, diameter 10cm
- bamboo canes or reeds, diameter 3–8mm
- silicone sealant (for aquariums) from DIY or pet shop
- secateurs

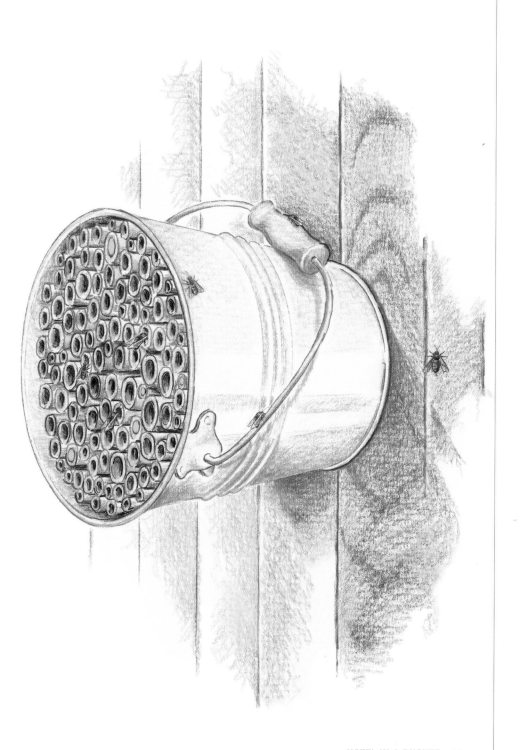

Reed hut

Time 3 hours
Build easy

Here you can attract a wide range of lodgers in a small space with a combination of reeds and bamboo as well as smaller and larger tunnels in wood.

1 First take the two side panels and cut the upper edges in a 45° angle using a jigsaw or circular saw.
2 Glue the two roof panels together at right angles and additionally secure them with 3 screws (pre-drill the screw holes).
3 Glue together the floor and side panels, as well as the roof, to create the finished frame and secure the connections with 3 screws each (again, pre-drill the screw holes).
4 Cut the upper part of the back panel to exactly match the outline of the house (see table below).
5 Paint the house and back panel with weatherproof paint if you wish.
6 Allow the paint to dry and then nail the back panel to the frame.
7 Cut several branches to a length of 120mm and drill suitable tunnels (page 9) into the cut surface.
8 Glue the branch pieces, the reeds or bamboo canes on to the back panel with wood glue.

Materials
- well-seasoned branch, eg from a fruit tree
- reeds and/or bamboo canes
- weatherproof, environmentally friendly paint
- wood glue
- rust-resistant screws 40mm
- rust-resistant nails 30mm
- jigsaw or circular saw, fine saw, drill, secateurs, cordless screwdriver

Quantity	Name	Material	Height/length (mm)	Width (mm)
1	floor	solid wood 18mm	114	120
2	side panels	see above	170	120
1	roof panel	see above	158	150
1	roof panel	see above	140	150
1	back panel	plywood 5mm	250	150

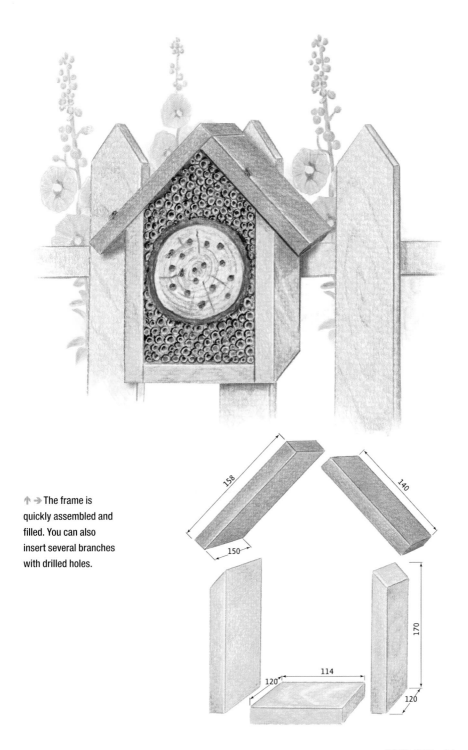

↑ → The frame is quickly assembled and filled. You can also insert several branches with drilled holes.

158

140

150

170

114

120

120

Nesting stone

Time 1 hour
Build very easy

You can buy nesting bricks but also easily make your own nesting stones using breeze blocks.

1 Cut your breeze block to the desired size using a jigsaw. You can also cut a triangle or hexagon in a honeycomb design: breeze block is very easy to work with.

2 Decide on your pattern and draw it on the breeze block. Remember to vary both the diameter and depth of tunnels (page 9). Most wild bees prefer diameters between 3mm and 6mm (and, accordingly, depths of 3–6cm depending on the diameter).

3 Drill tunnels into the breeze block and then carefully tap out the bore dust from the holes.

The home-made nesting stone should be hung or set up in a spot away from the weather, for example, on a shelf on a sheltered balcony.

Materials
- breeze block (aerated concrete block) from builders' merchant, 24.9 × 59.9 × 15cm
- jigsaw or circular saw, drill

Swedish shelf sculpture

Time 1 hour
Build very easy

The drilled sides of simple wooden shelves are perfect for easily built accommodation for wild bees.

1 The uprights of adjustable shelving units come with ready drilled holes for supporting the shelves, and are ideal for transforming into nests. Check the holes are clear of debris and remove any remaining pegs.

2 Cut the drilled battens to various lengths and glue them together in a design of your choosing.

3 Smooth the interior of the tunnels with a small round file. If the holes are not very deep, you can make them deeper with a drill.

4 Paint the exterior surfaces with environmentally friendly wood preservative or weatherproof, environmentally friendly paint; you can also paint each batten in a different colour before gluing them together.

5 Hang on a wall using screw eyes and sturdy string. Make sure to choose a sheltered spot protected from the weather on the balcony or patio wall. Done!

These shelves are found in many garages as left-over furniture from student or apprenticeship times. They're usually made of pine, which is not ideal, but still workable. Beech is a better choice.

Materials
- old sides of untreated wooden shelves with holes that are intended for pins supporting the shelves; also individually available at low prices from large furniture shops
- wood glue
- environmentally friendly wood preservative or environmentally friendly weatherproof paint
- saw, file, possibly drill

Rustic tree house 1

Time 1 hour
Build very easy

A classic design – simple, beautiful and popular. A nice drill pattern spruces the whole thing up.

1 It's important that the wooden cross-section is from a suitable tree: trunks of beech, oak, robinia (false acacia) or fruit trees are best.
2 The tree trunk must be well seasoned, or the cross-section will eventually develop radial cracks. Tunnels with such cracks running through them will not get colonised, so store freshly cut wood in a dry location for a while and then drill the tunnels between any cracks that have appeared in the wood.
3 A wooden cross-section for hanging on a wall should be about 12cm thick. Alternatively, you can use a thicker section of tree trunk and prop it up on a shelf or wall.
4 Drill holes in a design of your choosing. The tunnels should have a diameter of 2–8mm and each tunnel's length should be ten times its diameter (eg a 6mm tunnel should have a length of 6cm). Use the full range of diameters to provide a home for many different species (see also page 9).
5 Carefully tap out all shavings from the holes and file the tunnel entrance holes smooth.
6 Hang up the cross-section on a wall protected from the weather using screw eyes and sturdy string.

If you're not lucky enough to be able to source a suitable cross-section from your own or a neighbour's garden, you should be able to find one at a saw mill or firewood supplier.

Materials
- cross-section of oak, beech, robinia (false acacia) or fruit tree
- drill, round files

Rustic tree house 2

Time   1 hour
Build very easy

*Another version if you
have thick branches or
a thin trunk available.*

1 Choose a thick branch or thin tree trunk at least 20–30cm in
   diameter, with smooth bark, for this design – beech logs are
   particularly well suited. Smooth bark is important for easily
   spotting the insects going in and out of their tunnels; this would
   be much trickier to observe on rough oak bark. The wood must
   be well seasoned and dry.

2 Halve a log lengthwise with an axe or circular saw.

3 Drill tunnels of different sizes; tunnels must always be horizontal
   even if that means you're not drilling vertically into the trunk
   surface. Tunnel diameters should be 2–8mm and each tunnel's
   length should be ten times its diameter, meaning a 6mm tunnel
   should be 6cm long (see also page 9). Red mason bees
   particularly favour this size: they're found everywhere and are
   reliable colonisers, even in cities.

4 Carefully tap out all shavings from the holes after drilling.

5 File the tunnel entrance holes smooth to ensure they'll be
   readily accepted.

Job done! Hang up the hotel on a wall protected from the weather
using screw eyes and sturdy string.

Materials
- **thick branch or thinner trunk of beech or fruit tree with smooth
  bark if possible, diameter 20–30cm**
- **circular saw or axe, drill, round files**

Rustic tree house 3

Time 1 hour
Build very easy

A nice way of combining wooden cross-sections of various sizes.

1 Drill the wooden cross-sections as described on page 26. Use only seasoned wood as drying cracks (so-called seasoning checks) would render the tunnels uninhabitable.

2 Screw the cross-sections together on the back using rust-resistant fittings from the DIY shop.

3 Saw the boards for the roof so that they overlap the cross-sections by at least 5cm towards the front and a little on the sides. Mitre the roof boards with a jigsaw to make the gable.

4 Screw the roof boards together and then screw the roof on to the cross-sections.

You can either paint the roof in a colour, nail some roofing felt on to it or apply a double layer of reed cut from a reed screen.

Materials
- 3 cross-sections in different sizes, of oak, beech, robinia (false acacia) or fruit tree, thickness approx. 15cm
- 2 boards for the roof to match the size of the wood cross-sections, width approx. 20cm
- 2–3 metal fittings/brackets
- rust-resistant screws
- drill, cordless screwdriver, round files

Recycled garden tools

Time 1 hour
Build very easy

Do you have any old gardening tools that only collect dust in the shed or garage? Make use of their decorative value and provide for insects at the same time!

1 You can use any gardening tool with a thick wooden handle. Think about where they would look decorative in spots protected from the weather, for example on a patio wall.

2 Mount the gardening tools as you like – the main thing is that they should be firmly fixed and not swing.

3 Then drill tunnels of 3–6mm in diameter into the wooden handles. Here you cannot drill as deeply as would be ideal (page 9), but even so you're sure to attract grateful colonisers.

Job done! You could also drill tunnels into table legs, wooden fence posts and many other things; most wild bee species are not overly fussy.

Materials
- old gardening tools with a thick wooden handle
- rust-resistant screws for mounting the tools
- drill, cordless screwdriver

Wood block hotel

Time   2 hours
Build   easy

A little creativity and a simple wood block and garden reed screen turn into a unique design.

1   The measurements of the wood block can vary; below is an example of a suitable size.
2   Cut the wood block at the top using a circular saw to make a pitched roof shape.
3   Decide on your design and drill tunnels of 2–8mm in diameter, with each tunnel's length measuring ten times its diameter (page 9).
4   Be sure to file the tunnel entrance holes smoothly to improve the chances of them being accepted and carefully tap out all shavings and dust.
5   If you plan to hang up the hotel in a spot somewhat exposed to the weather, you should paint the wood block with an environmentally friendly wood preservative prior to affixing the roof.
6   To make the roof, cut a reed screen to size so that, once folded to make two layers, it overlaps the wood block by about 2cm at the front and sides.
7   Secure the cut and double-layered reed screen on the wood block with a few fencing staples – and your thatched roof is complete.

If the hotel hangs in a weather protected spot, even the thatched roof might get colonised!

Materials
- beech, oak or fruit tree wood block, eg 160 × 220 × 150mm
- garden reed screen
- rust-resistant fencing staples
- circular saw, drill, round files, secateurs

Alpine chalet

Time ½ day
Build easy

*Also fits on a city
balcony – a paradise for
mason bees, etc.*

1 Glue the two roof panels together at right angles and additionally secure them with 3 screws (pre-drill the screw holes).
2 Paint the roof in a colour of your choosing or nail roofing felt on to it.
3 Cut the upper edge of the roof support at a 45° angle using a jigsaw to ensure the roof attaches nicely later.
4 Drill holes in your chosen design into the two wood blocks. Vary the tunnel sizes and observe the correct ratio between diameter and depth (page 9) to provide nesting tunnels for many different species. Smooth down the cut edges and surfaces with sandpaper and a file.
5 Glue the wood blocks together as shown in the illustration.
6 Glue the roof support to the side of the higher wood block and secure it with 2 screws after pre-drilling the screw holes.

Materials
- environmentally friendly weatherproof paint or some roofing felt
- rust-resistant screws 40mm
- rust-resistant nails 30mm
- sandpaper, round files
- jigsaw or circular saw, drill, cordless screwdriver

Quantity	Name	Material	Height/length (mm)	Width (mm)
1	large block	solid wood 100mm thick	240	180
1	small block	see above	200	120
1	roof panel 1	solid wood 18mm thick	320	140
1	roof panel 2	see above	340	140
1	roof support	see above	120	100
1	back panel	plywood 5mm thick	400	300
many	filling material	bamboo canes or reeds	100	–

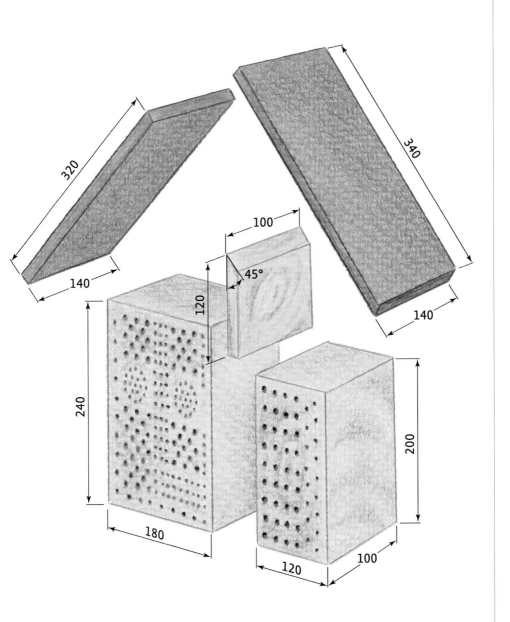

7 Glue the roof to the roof support and upper edges of the wood blocks. Secure it on the roof support and block edges with several rust-resistant screws (pre-drill the screw holes).

8 Lay down the hotel on the panel for the back panel (see table) and outline the roof angle. Cut the back panel with a jigsaw.

9 Affix the back panel to the wood blocks and roof with some rust-resistant nails.

10 Fill the roof space with appropriately cut bamboo canes or reeds (page 9). You can also mix the bamboo canes or reeds with 2–3cm thick, well-seasoned branch logs into which you first drill tunnels as you did into the wood blocks.

11 Secure the bamboo canes, reeds and/or branches with wood glue or silicone as described for the 'Hotel in a Bucket' (page 18).

All done! Now you can hang up your hotel in a sheltered spot on your balcony. You'll be surprised how popular this new accommodation will prove, even on a city balcony.

Tile studio apartments

Time 1 hour
Build very easy

It can't get any easier: perforated bricks such as extruded interlocking tiles only need polishing and stacking.

1 Extruded interlocking tiles are rarely found as they are no longer commonly used. Try recycling companies specialising in old construction materials, or demolition companies. Purchasing extruded interlocking tiles online is usually tied to minimum order quantities. Alternatively, you can use perforated bricks with small holes.

2 Halve the long tiles with a suitable saw.

3 The openings of these tiles and of perforated bricks are rarely smooth. Use a round file to deburr the openings so that insects will accept such tiles and bricks.

4 As the tunnels are open on both sides, they must be closed at the back with some clay from a crafts shop.

5 Stack the bricks and place them on a shelf on the balcony or patio close to a wall. They should be protected by a roof. The bricks look nice scattered between herb and other flower pots. Don't put any pots without saucers on top of the bricks as they'll get too wet. The bug hotel is ready!

It's important to leave the bricks in place over the winter. If you need to remove the flower pots, leave the stacked bricks where they are. The larvae of wild bees and solitary wasps need this hibernation, with corresponding outdoor temperatures.

Materials
- extruded interlocking tiles or perforated bricks with small holes
- round file
- clay (crafts shop)
- masonry saw

Observation cabinet 1

Time 3 hours
Build easy

Even if you don't have a woodworking router, you can easily build an observation bug hotel using a small wall-mounted cabinet from a furniture shop.

1 Saw the battens into pieces of about 10cm in length, depending on the depth of your cabinet. Use a saw guide to ensure the cuts are at right angles.

2 Clamp the battens together in pairs flush with each other. You can make the tunnels (closed at one end) as follows: drill off-centre first into one batten, then into the other so that the drill holes overlap in the middle (see illustration). This provides relatively round tunnels which you'll be able to look into (see also page 9).

3 Smooth the tunnels with a round file and the entrances with sandpaper.

4 Using a jigsaw, carefully saw the acrylic glass into pieces of the same dimensions as the broad sides of the battens with the tunnels or have them cut to size directly at the DIY shop. Use a hot glue gun to glue the acrylic glass panes on to the battens (obviously avoiding glue getting into the tunnels). Alternatively, you can fasten these window panes with elastic bands.

5 Drill entrance holes of 3–8mm in diameter into the door of your cabinet; they must be the same diameter as the tunnels they lead to.

6 Secure the matching window batten behind each hole by screwing on metal angle brackets so that the insects can directly enter the tunnels through the holes in the door. Ensure that no light penetrates through the edge of the acrylic glass.

Materials
- wall-mounted, lightproof wooden cabinet (eg medicine cabinet), depth 12–15cm
- wooden battens 20 × 30mm
- acrylic glass 3 or 4mm thick
- rust-resistant screws, small metal angle brackets
- hot glue gun or sturdy elastic bands
- standard wood saw, saw guide, screw clamps, drill, jigsaw, round file

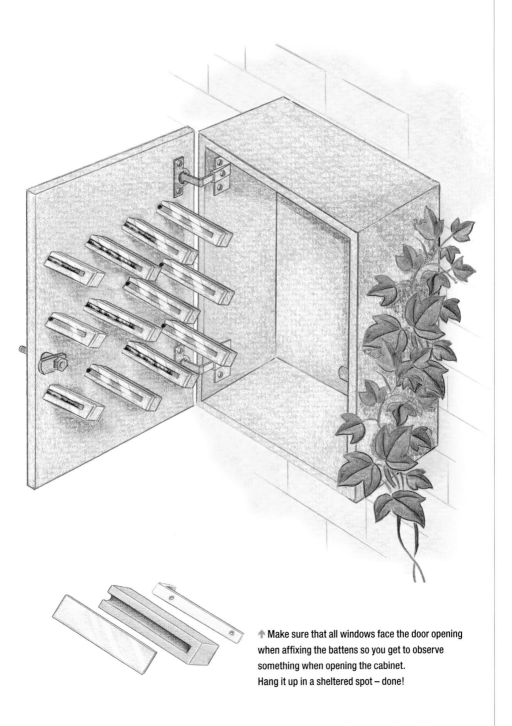

⬆ Make sure that all windows face the door opening when affixing the battens so you get to observe something when opening the cabinet.
Hang it up in a sheltered spot – done!

Observation cabinet 2

Time ½ day
Build moderately
difficult

*Exciting: this hotel, too,
lets you observe what
the next generation is
up to.*

1 Cut tunnels of various diameters and lengths into the thicker nesting panel at intervals of about 20mm – this will attract different species that also build different nests. The tunnels should measure 3, 4, 5, 6, 7 and 8mm in diameter and each tunnel's length should be ten times its diameter, that is, 30, 40, 50, 60, 70 and 80mm.

2 Saw the batten to build a frame: you'll need two 120mm long pieces and two 320mm long pieces.

3 Glue the frame to the nesting board and make sure that your acrylic glass panel fits inside the frame. Should you want to additionally secure the frame, use rust-resistant nails with flat heads or countersunk screws: the surface must be level.

4 Put the acrylic glass pane into place and, without too much pressure, pre-drill the screw holes with a cordless drill. The screw holes should have the same diameter as the screws because acrylic glass cracks easily.

5 Screw on the acrylic glass window.

6 Affix the lid with hinges, then the hooks and screw eyes for securing the lid when it's shut.

Now you can hang up the hotel in a sheltered spot away from the rain. When open, the window should not be subject to direct sunlight. Make sure you always close the lid securely so that it stays dark inside

Materials
- 2 rust-resistant hinges of matching size with screws
- 2 rust-resistant hooks and screw eyes to close the lid
- rust-resistant screws, about 15mm in length
- hand-held router, wood saw, cordless drill

Quantity	Name	Material	Height/length (mm)	Width (mm)
1	nesting board	glued wood beech 21mm	320	150
1	lid	glued wood beech 15mm	320	150
1	batten for frame	square batten pine/spruce 5 × 15mm	1m (880mm are required)	
1	window	acrylic glass 5mm thick	290	120

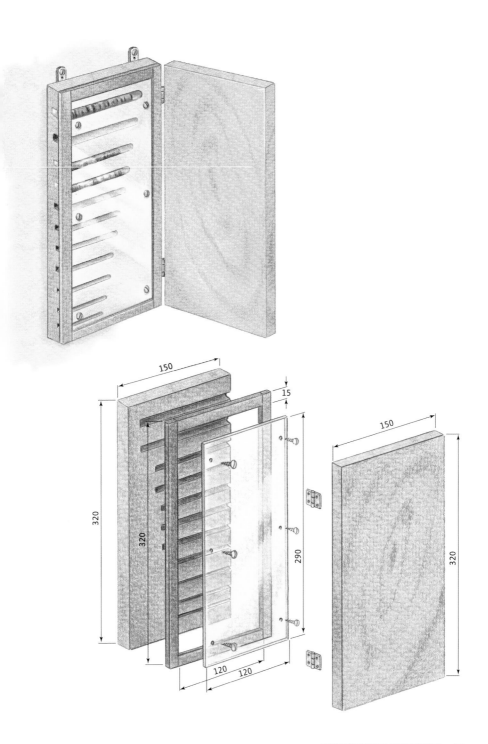

150

15

150

320

320

290

120   120

320

Mud commune

Time 1 hour plus
time for
drying
Build very easy

*Mud is a rare
commodity in many
gardens, yet many wild
bees dig their nests in
it. Others use it to build
walls and 'front doors'
for their breeding cells.*

1 Gradually fill the planter by the handful with your loam or clay mixed with sand.
2 Compact the mud carefully (especially when using a terracotta planter) to eliminate any air pockets.
3 Use a thin stick to bore only a few holes of about 6–8mm in diameter and 1–2cm in depth into the mud. These holes are not nesting tunnels but will make it easier for the insects to access the inside of the mud block. The insects will chew their own nesting tunnels into the mud.

After the mud has fully hardened, place the planter horizontally on a protected shelf away from the weather (eg between your herb pots) so that the 'mud wall' is vertical.

> **Materials**
> - an untreated ceramic planter, about 40cm long
> - mud: loam or clay, mixed with sand, see page 12
> - thick branch or log for compacting

Winemaker's house

Time   3 hours

Build   very easy

A nice combination for mud dwellers, mud builders and tunnel-nesting insects and very easy to make.

1 Gradually fill 4 of the 6 round holes of the terracotta wine rack with mud and compact it with a log of wood to eliminate all air pockets. A few holes like in the 'Mud commune' model (page 46) to make access easier for the insects.

2 Put the wine rack on a smooth, clean surface with the holes facing down and leave it in this position until the mud has hardened.

3 Cut reeds or bamboo canes, as described on page 9, so that the stalks or canes roughly measure the wine rack's depth. You can also combine reeds and bamboo canes. Fill the cut reeds or bamboo canes into the remaining 2 holes of the wine rack so that they are firmly stuck in the holes.

Job done! Place the wine rack, with the holes horizontal, in a sheltered, sunny spot on a patio or balcony.

Materials

- terracotta wine rack
- mud: loam or clay, mixed with sand (see page 12)
- bamboo canes and/or reeds (page 9)
- thick branch or log for compacting
- secateurs and/or fine saw

Bee borders

Time 1 weekend
Build moderately
complex

Are you planning to build a small raised bed or rock garden? Then you could design the border to provide a fantastic hotel for solitary bees and wasps.

1 The number of hollow-core retaining blocks and wooden beams you use depends on the size of your raised bed; however, you can choose to design only part of the border in this manner.

2 Place the hollow-core blocks with the opening facing upwards on a smooth surface and compact the mud after filling (see page 46). Set aside to allow the mud to harden.

3 For the wooden elements between the hollow-core retaining blocks, saw the beams into equal-length pieces; the actual length is down to your choice and design.

4 Connect the beams by screwing the wooden battens to the back. The battens should protrude about 20–30cm at the bottom. This will allow you to secure the wooden beam elements in the ground to stop them toppling over.

5 Design your raised bed border using 2 or 3 stacked, mud-filled retaining blocks and intermediate wooden beam elements. The retaining blocks and wooden beam elements should be of the same height.

6 Finally, drill tunnels into the wooden beams (see page 9).

In combination with the 'Guest house for diggers' on page 64, you now have created a perfect paradise for wild bees and wasps.

Materials
- wooden beams, about 15 × 15cm (suitable wood: see page 8)
- rectangular, stackable hollow-core retaining wall blocks, about 20 × 40 × 25cm
- square wooden battens about 2 × 2cm for connecting and securing the wooden beams
- mud: clay or loam (page 12)
- rust-resistant screws
- jigsaw, drill, cordless screwdriver, round files

High rise hotel

Time ½ day
Build very easy

This hotel is quickly filled and ideal for all who have only limited space on their patio.

1 If your CD shelf is made of painted fibreboard (MDF), it should be placed in a sheltered spot away from the rain. CD shelving units made of solid wood are preferable as they are more robust.

2 If the shelving unit is made of wood, you can paint the exterior with environmentally friendly protective paint, if you wish.

3 Start by filling one or two sections of the shelving unit (lay it down on the floor) with mud as described on page 46 and compact the mud.

4 Cut reeds and/or bamboo canes as per instructions on page 9 and fill further sections with these. It's advisable to glue the reeds or bamboo canes to the back panel with wood glue. If necessary, protect the reeds or bamboo canes from hungry birds by affixing galvanised wire mesh on the front.

5 Fill the remaining sections with logs and thick branches or breeze blocks cut to size that you have previously drilled tunnels into.

Place the slimline hotel in a sunny, sheltered spot. You might want to secure it to a wall by drilling through the back panel or using metal angle brackets to prevent it from falling over.

Materials
- CD shelving unit, made of solid wood if possible
- reeds, bamboo canes
- mud: clay or loam (page 12)
- thick branches or wood blocks
- breeze block
- saw, drill

Honeymoon villa

Time 1 day
Build moderately
complex

Materials
- reeds, bamboo canes
- garden reed screen for thatched roof
- mud: clay or loam (page 12)
- thick branches or wood blocks
- 1 small piece of galvanised wire mesh
- wood glue
- rust-resistant screws 40mm
- rust-resistant nails 20mm and fencing staples
- saw, drill, cordless screwdriver, stapler if necessary

1 Glue together the floor, side panels, intermediate floor and partition walls, making sure that everything fits together well: the intermediate floor and partition walls are offset to the front by 6mm compared to the frame so that the back panel (see table) can be inserted later.

2 Pre-drill screw holes with a thin drill bit and then secure everything with screws. You can choose to screw it all together straight away without previously gluing, but that is a bit more difficult.

3 Then screw the roof on to the hotel. It should be flush with the side panels at the back; towards the sides and front it protrudes by a few centimetres for weather protection.

4 Now nail the back panel to the frame (consisting of side panels, floor and roof) and to the partition walls.

5 Fill the hotel with reeds or bamboo canes (page 9), wood with drilled tunnels (page 8) and mud (page 12). You should glue the reeds and bamboo canes to the back panel or secure them with silicone (page 18).

6 Cut the 'balcony railing' to size and affix it with a few fencing staples or a stapler.

7 Thatch the roof with 2 layers of garden reed screen nailed down with fencing staples.

As a final touch you could add a hotel sign made of leftover plywood at the centre of the intermediate floor.

Quantity	Name	Material	Height/length (mm)	Width (mm)
1	floor	solid or glued wood beech 18mm	500	130
1	intermediate floor	see above	464	124
2	side panels	see above	300	120
1	roof	see above	600	170
5	partition walls	see above	141	114
1	back panel	plywood 6mm thick	464	300

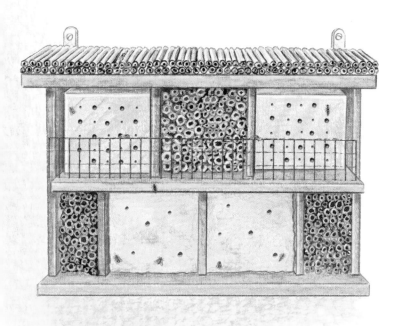

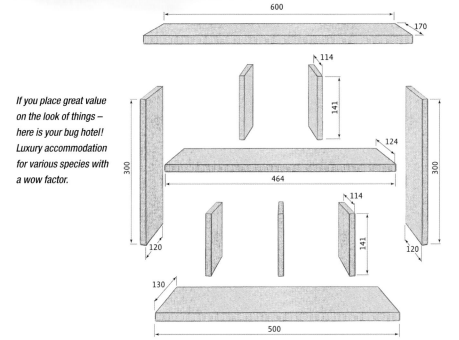

If you place great value on the look of things – here is your bug hotel! Luxury accommodation for various species with a wow factor.

Swedish cabin

Time 1 day

Build moderately complex

Materials
- reeds, bamboo canes
- mud: clay or loam (page 10)
- thick branches or wood blocks
- wood glue
- stainless steel screws 40 mm
- environ-mentally friendly weatherproof paint
- saw, drill, cordless screwdriver, chisel or flat-head screwdriver, hammer, brush

1 Start by sawing the partition wall to size.

2 Then cut the upper edges of the side panels and the connecting edges of the roof panels in a 65° angle using a jigsaw. Cut the upper edges of the vertical partition wall in a pitched roof shape to create a 130° angle at the apex.

3 Pre-drill all screw holes and screw together the floor, side panels and partition wall; it's easier if you glue them together first.

4 Then affix the roof with glue and screw it to the partition wall and side panels.

5 Saw twice 18mm apart halfway down into the vertical and horizontal intermediate partitions as shown in the illustration and chisel out the bit of wood between the cuts with a chisel or flat-head screwdriver and hammer. Now you can slot the partitions together and glue them at right angles. The best way to do this is to insert them straight away into the frame of the house, align them and glue them into the frame.

6 Fill the hotel with reeds or bamboo canes (page 9), wood with drilled tunnels (page 8) and mud (page 12). You should attach the reeds and bamboo canes on the back panel with wood glue or silicone (page 18). If you want to paint wood featuring drilled tunnels, make sure that no paint gets into the tunnels. It's best to leave a small circle of bare wood around the tunnel holes.

To make a real feature of the hotel, mount it on a strong fence post free-standing in the garden. Affix the house using rust-resistant metal angle brackets and drive the fence post at least 50cm deep into the ground to make it stand firm.

Quantity	Name	Material	Height/length (mm)	Width (mm)
1	floor	solid or glued wood beech 18mm	400	260
2	side panels	see above	230	260
2	roof	see above	320	340
1	partition wall	see above	364	364
2	intermediate partitions horizontal	see above	364	121
2	intermediate partitions vertical	see above	314	121

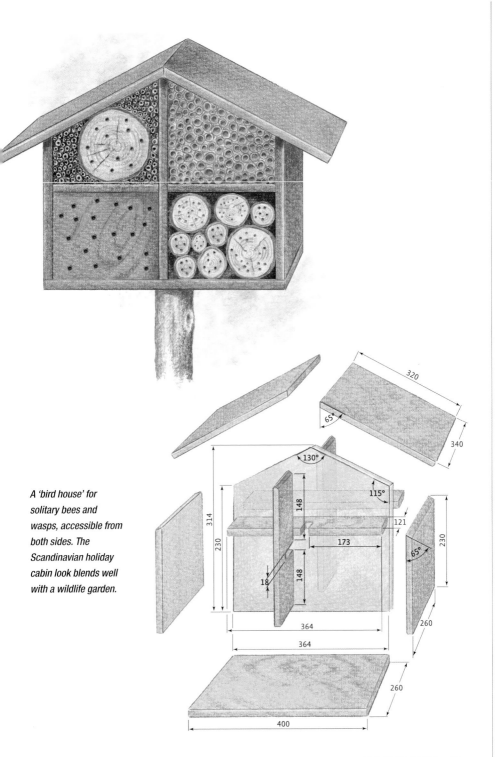

A 'bird house' for solitary bees and wasps, accessible from both sides. The Scandinavian holiday cabin look blends well with a wildlife garden.

320

65°

340

130°

115°

314

230

148

121

173

18

148

364

364

260

260

65°

230

260

400

Weatherproof hotel

Time 1 day
Build moderately complex

1 The observation box that is part of this design is simply slotted into the narrow frame. See illustration for construction details. For guidance on how to cut the tunnels, follow the instructions on page 42.

Materials

- reeds, bamboo canes
- garden reed screen
- roofing felt
- breeze block for sawing to size
- mud: clay or loam (page 12)
- thick wood blocks
- wood glue
- stainless steel screws 40mm
- rust-resistant nails 10mm and fencing staples
- router, saw, drill, cordless screwdriver, brush

2 Start by sawing the roof shape into the back panel: the roof apex has a right angle. The back panel simply gets screwed to the back at the end.

3 Now screw together the floor, side panels, intermediate floors and supporting wall for the observation box. Make sure that everything is at right angles. Then screw on the roof and finally the back panel.

4 Choose from a wide range of fillings for your hotel. One option: cut tunnels of 3–8mm in diameter that are closed at the end into MDF boards, which you then stack one on top of the other (see book cover). The picture here features breeze block that was cut to size, bamboo canes, wood blocks and thick branches as well as a mud-filled compartment.

5 Cut the roofing felt to size and nail it to the roof. To improve the overall look, you can add a nice thatched roof using a garden reed screen affixed with fencing staples.

Hang up the hotel on a wall or mount it on a fence post, facing south-east. It does not require additional rain cover. This hotel is particularly weather proof and durable. Textured coated board is more expensive than blockboard, but the investment pays off.

Qty	Name	Material	Height/ length (mm)	Width (mm)
1	floor	textured coated board 15mm	370	140
2	side panels	see above	450	140
2	intermediate floors	see above	370	140
1	supporting wall on observation box	see above	200	140
1	roof panel	see above	350	170
1	roof panel	see above	335	170
1	back panel	textured coated board 12mm	670	400
1	observation box: board for cutting tunnels into	glued wood beech 18mm	198	135
1	observation window	acrylic glass 4mm	198	135
1	batten for tunnel entrance holes	plywood 5mm	230	40

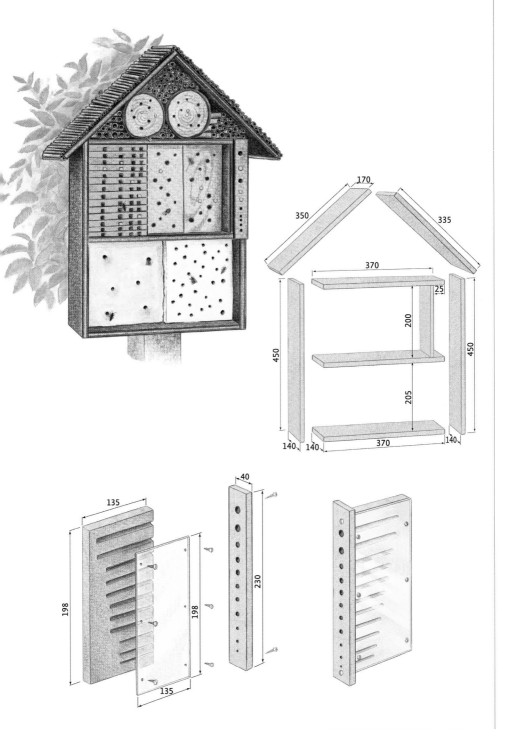

Grand hotel

Time 1 weekend
Build complex

This hotel is a great project for nursery schools, school gardens or a joint neighbourhood activity: everyone can add their ideas as there is plenty of space available.

1 First taper the upper edges of the side beams with a panel saw so that the gable will have an angle of 90°. Smooth the cut surfaces with a rough file/rasp.

2 Assemble the basic frame by screwing the roof on to the beams and affixing the shelves on the side beams with metal angle brackets.

3 Connect the shelves with partition walls, cut to size, again using rust-resistant metal angle brackets. That will provide additional stability for the large hotel.

Materials

- 2 post spikes/ground screws in the right size
- reeds and bamboo canes, empty food tins to hold them
- wood blocks, logs, wood cross-sections
- extruded interlocking tiles or perforated bricks with small holes
- breeze block
- terracotta wine rack
- green lacewing box (page 88)
- mud: clay or loam
- roofing felt
- rust-resistant metal angle brackets
- rust-resistant screws
- panel saw, jigsaw, drill, cordless screwdriver, rough file/rasp (flat), round files

Quantity	Name	Material	Height/length (mm)	Width (mm)
2	side beams	solid wood 150 × 150mm	1500	
4 or 5	shelves	solid wood 27mm thick	1000	150
about 7	partition walls	see above	variable	150
1	roof	see above	1500	327
1	roof	see above	1500	300
1	roof panel	see above	335	170

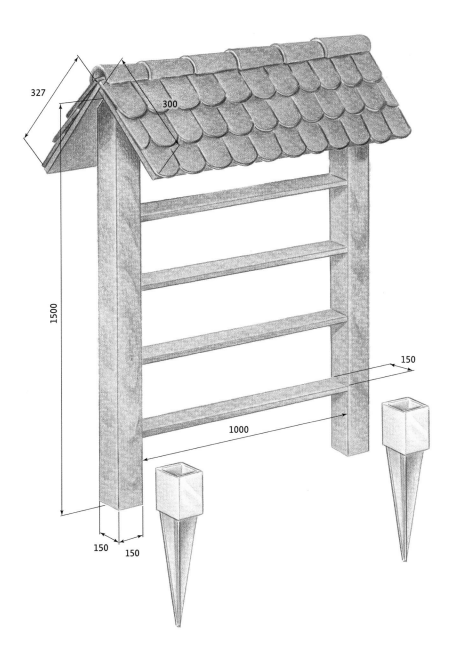

4 Secure the hotel in the ground using post spikes. Depending on firmness of the ground you may need to compact it first: dig out soil to about a quarter of the post spike's depth and a little more than its width where you intend to drive it in. Then fill the hole with gravel or frost-protection aggregate and ideally compact it with a plate compactor (hire from builders' merchant). For very large bug hotels you should set the posts in frost-resistant concrete because of the wind load. The hotel should be set up in a sunny spot with the nesting side facing south or south-east.

5 Nail roofing felt on to the roof. You can make it look prettier with garden reed screen or roof it with roof tiles (more difficult, but better-looking version).

6 Now you can fill the shelves. You have a wide range of options available: fill empty food tins with reeds or bamboo canes (best glued to the bottom of the tin to prevent destruction by hungry birds). If you have perforated bricks as one often sees in large bug hotels, fill the holes with reeds – otherwise the bricks will be useless as their holes are too big. You can also include wood blocks, logs and thick branches with drilled tunnels. Observe the correct diameters and depths (page 9) when drilling the tunnels. You can also drill into breeze blocks. Extruded interlocking tiles with ready-made tunnels are easily stacked on the shelves (more on these tiles on page 40). It's very important to provide mud for species that nest in this material and those that need it for their nest's partition walls and nest seals. You can put mud in terracotta flower pots, wine racks or similar receptacles (more information on page 12).

7 One or several green lacewing boxes, painted red, look good and include yet another species. Of course, you can also integrate a butterfly shelter (page 92) into the grand hotel – ideally without a pointed roof.

You don't need to fully fill the bug hotel straightaway – regular addition of new features from April to August provides new attractions for many species.

Guest house for diggers

Time   1 hour
Build   very easy

Most solitary bees nest in the ground, but the insects don't always find a suitable location. With only a few simple means you can provide them with optimal conditions on your balcony.

1. You have two options for the soil: ask for unfertilised or only moderately fertilised cactus potting compost at your local garden centre. If you can't get this, mix your own soil as follows: 1 part loam, 1 part fine sand and 1 part seeding compost. This provides soil that is neither too solid nor too loose. The insects must be able to easily dig tunnels without the soil crumbling down on them.

2. Put some terracotta shards on the bottom of the terracotta pot for good drainage.

3. Fill up with ground and plant the sedum; however, it should occupy only a small part of the pot.

4. Put some stones on top of the soil for decoration, but make sure that a lot of soil remains visible.

5. Water well to set the soil.

6. Place the pot in a sunny, sheltered spot on your balcony or patio. It should remain weed-free. Every now and then, water moderately.

Don't despair if your pot isn't instantly colonised by insects – this can take a while. In the meantime, the insects appreciate the nectar-rich sedum flowers.

Materials
- terracotta pot of about 30cm or more in diameter, with drainage hole, at least 15cm deep
- suitable soil (see description)
- stones and sedum for decoration

Diggers' rock garden

Time ½ day
Build very easy

A rock garden in a raised bed looks good and doesn't require much space in the garden. You'll be providing a valuable habitat for ground-nesting wild bees.

1 Select a very sunny spot in your garden. You should build a small wall to make a raised bed for your rock garden (eg the one on page 50 or a drystone wall) – the soil will benefit from improved drainage.

2 You could also design the bed to form a small herb spiral and raise it that way.

3 Mix 1 part loam, 1 part sand and 1 part potting compost; the total quantity depends on the size and depth of your raised bed. You should have a layer of at least 20cm of this special soil.

4 Plant some scattered flowering rock garden plants with plenty of bare soil between them and choose slow-growing varieties such as sedum, house leek (*Sempervivum*) and low-growing, unfilled bellflower (*Campanula*).

Keep the bare patches of soil free from weeds and your rock garden will soon attract ground-nesting wild bees.

Materials
- suitable soil (see description)
- stones and slow-growing, flowering rock garden plants

Live-in sculpture

Time ½ hour
Build very easy

*Do you or your
neighbour have an old
dead fruit tree? Perfect
for the insect world!*

1 Dead wood from fruit trees is a veritable paradise for many
solitary bees that chew their own tunnels into the decaying
wood. It also adds the beauty of a natural sculpture to your
garden.

2 Use the trunk or thick branch (it should have a diameter of at
least 20cm) by placing it in a sunny spot on the patio or leaning
it against a wall. You could even secure an entire trunk with
hooks and wire cable upright against a wall.

3 This will gradually attract a range of insects that bore their own
tunnels into the wood or use other insects' tunnels for nests.

4 Of course, you can also pre-drill some tunnels (page 8) to
provide a range of options from the start.

With a little luck you might even get to observe the huge but
harmless violet carpenter bee on large, upright trunks. You can
take further steps to attract it by planting wisteria close by – it loves
the blossoms!

Materials
- **trunk or thick branch from an old fruit tree**
- **drill, hook and wire cable if necessary**

Chapter 2
Hotels for large families

Bumblebees and hornets

Bumblebees have a fuzzy, teddy bear appearance that helps them achieve the reputation of peaceful, beneficial insect. With a far more alarming presence, hornets have a harder time, although there are plenty of reasons to offer a home for this large wasp too.

⬆ Diligent pollen collector – a buff-tailed bumblebee.

In contrast to a solitary bee, bumblebees live socially in small summer colonies with up to several hundred workers, established by a single queen between March and June.

Most species seek out deserted rodent nests as a nesting site, for example, in compost heaps and underneath piles of dead wood. Others move into empty bird nest boxes or loft insulation. The queen broods and raises the first generation of six to eight workers, covering her eggs with wax. Once the first generation emerge as adults, the queen can remain in the nest. The workers forage for pollen and nectar, build honey and pollen pots and care for the eggs and larvae. In summer, males and new queens are raised that mate outside the nest. While the young queens start looking for winter quarters as early as in August, the workers and old queen gradually die.

Bumblebee castles

Bumblebee hotels are no substitute for a wildlife garden with 'wild corners' but offer a good opportunity for observing these insects. You may need a lot of patience before these hotels are colonised. Only some species, such as the red-tailed bumblebee, garden bumblebee and tree bumblebee, return to the old nesting site year after year. Bumblebees in bumblebee boxes do require care: these kinds of nest are often invaded by parasitic wax moths. This can be prevented by installing a wax moth flap in front of the entrance hole which the bumblebees can open. The interior furnishings should be replaced during the winter season.

Gentle giants

Why should you provide homes for hornets? By doing so, you'll be helping a species that is under threat in many places. European hornets will repay you by hunting wasps, and eating the wax moths that damage bee colonies. Invasive Asian hornets, however, attack bees, and should not be welcomed. Hornets may appear formidable due to their size, but their sting is no more dangerous than that of a wasp, and they only sting when under attack. They remain perfectly calm, for example, while you mow the lawn or play ball.

The life cycle of hornets is like that of bumblebees: they, too, live socially in colonies of several hundred workers that form in May and die in October. They

Caution: Sting!

If you want to peek inside the nest of a large bumblebee colony, you may want to wear protective clothing – bumblebees can sting.

build castles up to 60cm long out of chewed-up wood in tree cavities, often also in bird nest boxes and even in lofts and wall cavities.

Hotels for hornets

Hornet nest boxes should be hung up in a sunny spot with free air space in front of the entrance. It's important to choose a quiet location high above ground where neither neighbours nor passers-by might feel disturbed. It may take a while for the boxes to be colonised; but hornet boxes can attract occupants as late as August if a colony has to move from a smaller nest. Gluing in old wasp or hornet nests makes the box more attractive.

⬇ Hornets love windfall fruit.

Bumblebee castle

Time ½ day
Build moderately
complex

*Your own bumblebee
colony in your garden is
quite a special thing,
but you may need to be
patient before a queen
chooses your box as a
home.*

1 Take the front panel and make an entrance hole measuring
2–3cm in diameter. You can do this by drilling several holes in a
circle, breaking out the perforation and finally smoothing it all
down with a file.
2 Screw together the entrance frame and affix it to the front
panel with screws. The lower board of the entrance frame
should be at the same height as the entrance hole so that the
insects can easily crawl inside.
3 Glue together the side, front and back panels to form a box.
4 Glue down the two battens on the floor, which later will protect
the cardboard from moisture.
5 Glue the floor to the box.

Materials

- roofing felt for the roof, about 52 × 52cm
- 2 sturdy cardboard boxes, 1 about 35 × 35 × 35cm;
  1 about 20 × 20 × 20cm, the latter open at the bottom
- cardboard tube, diameter 2–3cm, length about 15cm
- environmentally friendly weatherproof varnish
- 60 stainless steel screws 30mm
- rust-resistant nails for the roofing felt
- small insect bedding/litter (eg wood shavings)
- filling material (see text)
- jigsaw, drill, hammer

Quantity	Name	Material	Height/ length (mm)	Width (mm)
1	roof	solid wood 18mm thick	440	440
2	roof battens	see above	360	20
2	side panels	see above	400	400
2	front/back panels	see above	400	364
1	floor	see above	400	400
2	floor battens	see above	300	30
4	feet	see above	40	40
2	entrance	see above	50	40
2	entrance	see above	100	40

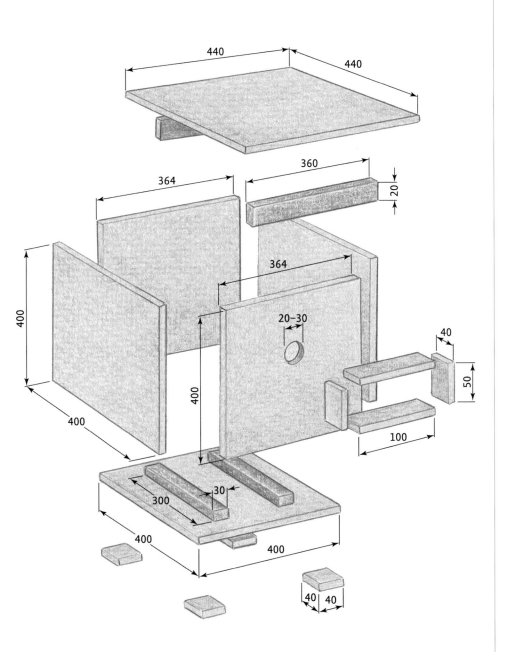

440

440

364

360

20

364

20–30

40

50

100

400

400

400

400

30

300

400

400

40 40

6 Now pre-drill the screw holes and screw together the side panels and floor.

7 Screw on the two battens for locking the roof/lid in place so that they fit just inside the side panels.

8 Screw the four feet on to the underside of the box to protect it against moisture from below.

9 Protect the roof and box from rain with a coat of environmentally friendly weatherproof varnish. Dark green blends in well with the surrounding garden.

10 Additionally, protect the roof with roofing felt; fold the felt over the edges and nail it down from below.

11 Cut a round opening of 2–3cm (depending on the diameter of the cardboard tube) into the large cardboard box and a small gate of the same size into the small cardboard box.

12 Put some small insect bedding/litter into the large cardboard box. Loosely fill the small cardboard box with fine hay, dry moss, kapok or abandoned, torn-apart tit nests from bird nest boxes.

13 Now place the large cardboard box inside the wooden box and put the small cardboard box into the large one with the opening facing down. If the small cardboard box is not fully closed at the top, firmly seal the flaps with adhesive tape.

14 Use adhesive tape to affix the cardboard tube as an access tunnel between the entrance hole and the living quarters.

15 You can put your bumblebee castle down on the ground in a semi-shaded spot underneath deciduous shrubs as of early March. It should be well protected from the full midday sun.

You'll have to carefully remove the small cardboard box as soon as the first workers emerge. Then the colony will have enough space.

Hornet hostel

Time 1 day
Build advanced

The queens prefer colonising smaller spaces in spring, so the box features an inserted shelf.

1 Start with the side panels. Taper them at the top as shown in the illustration.
2 Cut the upper edges of the batten above the door and the back panel into a 70° angle using a jigsaw or circular saw. Cut one of the 250mm edges of each of the two floor panels to form the same 70° angle.
3 Saw the entrance hole of 1.5cm width and 6cm height into the door. The upper edge of the entrance should be 8–9cm below the upper edge of the door.
4 Screw on the broader floor panel at an angle at the very bottom of the side panels and the shorter floor panel at an angle at the very bottom of the back panel and side panels. Once the box is closed, a gap of about 1.5cm should remain between the two floor panels.

Materials
- roofing felt or a piece of pond liner to line floor and roof
- 1 twig or dowel rod, diameter about 1cm, length 270mm, as a safety rod
- rust-resistant screws 40mm
- thick, sturdy wire and a piece of old garden hose for hanging up the box
- jigsaw, drill, hammer, cordless screwdriver

Quantity	Name	Material	Height/length (mm)	Width (mm)
1	front floor panel	spruce 18mm thick	165	250
1	back floor panel	see above	135	250
2	side panels	see above	670	270
1	door (front panel)	see above	480	286
1	batten above door	see above	50	286
1	back panel	see above	600	250
1	roof	see above	390	340
1	landing platform	plywood 6mm thick	100	40
1	inserted shelf	see above	258	250
4	side battens	wooden battens 20 × 20mm	258	
1	nesting batten	rough batten 30 × 10mm	260	
1	crawl batten	see above	about 135	

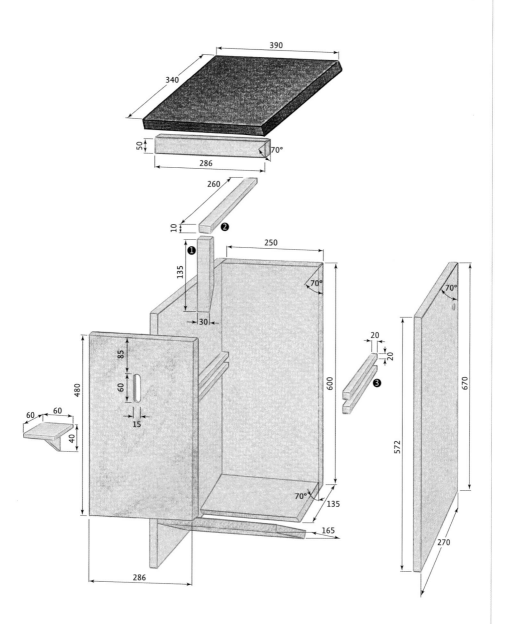

5   Line both floor panels with roofing felt or a thick, sturdy plastic liner on the inside of the box so that the lining extends a bit up along the walls and overlaps the bottom edge. Affix it with rust-resistant nails. Hornet faeces is liquid and would otherwise quickly soften the wood.

6   Screw together the side panels, back panel and roof, and screw on the batten above the door on the front as a spacer between sloping roof and door – otherwise the door cannot be opened.

7   Screw two side battens, 7mm apart, on to each of the side walls just below the level of the lower edge of the entrance hole. This is where the shelf will be inserted; you need to be able to remove it at any time.

8   Cut the lower end of the 'crawl batten' to make a 45° angle. This batten must be very rough, just like the 'nesting batten', or else the hornet queen will have difficulty climbing up on the smooth inside walls. If you're making the box out of wood that provides a rough, unplaned inner side to start with, you can do without the battens.

9   Screw the 'nesting batten' on to the box ceiling and the 'crawl batten' from the inside on to the batten above the door; the lower, tapered end should be exactly level with the upper edge of the entrance hole.

10 Later you'll need a safety rod to replace the inserted shelf: a stable stick or a dowel rod of about 1cm in diameter. Saw or file the safety rod to size so you can later insert it firmly between the side battens. This will help to stabilise the nest and the hornets will simply incorporate the rod.

11 For the landing platform saw the plywood into a 4 x 6cm piece and the remaining square into two identical triangles. Glue the platform on to the door directly underneath the entrance hole. Attach the door with two hinges and affix a hook and screw eye for closing it.

12 Drill two holes left and right into the very top of the back panel that exactly match the diameter of your wire for hanging up the box. That is much more secure for the heavy box than battens or metal screw eyes.

13 Paint the exterior of the box with environmentally friendly wood preservative or paint.

14 Nail roofing felt or pond liner on to the roof, overlapping the edges, to protect it from rain.

15 Thread thick, sturdy wire through the holes in the back panel.

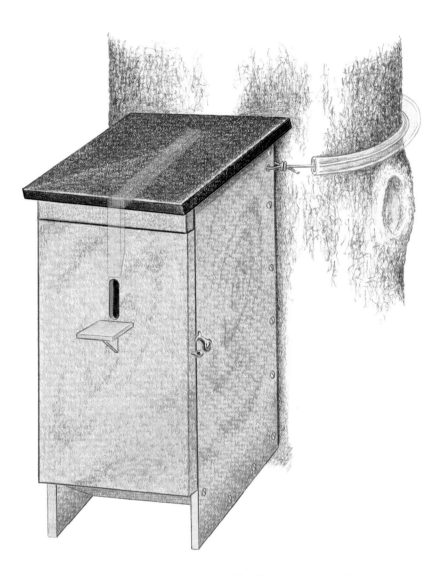

⬆ Hang the box on a tree trunk by putting the wire around the trunk and don't forget to first thread it through a piece of garden hose as a sleeve to avoid it growing into the trunk, damaging the tree. The box should face south-east. Once nest building has begun, you'll have to remove the inserted shelf (while the queen is out foraging) and insert the safety rod instead as described in the instructions.

Chapter 3
Hotels for beneficial bugs

Lacewings, ladybirds, butterflies

Bug hotels offer a home not only for bees and wasps but also for other insects: green lacewings, ladybirds and others will reward you by eagerly devouring aphids, helping you maintain your garden without pesticides.

↑ An aphid lion – the green lacewing larva – is a voracious predator of aphids and whiteflies.

Graceful lacewings

Green lacewings are bright green net-winged insects. They are harmless, somewhat awkwardly flying insects that are rather pretty on closer examination. They lay their eggs in the proximity of aphid colonies. By securing their eggs at the ends of long silk stalks they protect these from ants guarding the aphids. The larvae, aptly named 'aphid lions', have a voracious appetite for aphids, mites, caterpillars and other insect larvae. You can attract them to the garden or balcony by offering them a protective day or winter home.

Suitable accommodation often also attracts earwigs. These insects hide in the daytime in large family groups and emerge at night to hunt. They are rather unselective predators and will eat aphids as well as plant parts.

Decorative bugs

Ladybirds, too, appreciate overnight or winter accommodation. Some 46 species of various colours can be found in Britain, and even more across Europe, although not all are recognisable as ladybirds. Ladybirds dine on up to 100 aphids, spider mites, scale insects and other pests – per day! Their offspring are equally voracious and therefore highly valued as plant protectors.

These days, however, competition from Asia introduced to Europe by humans threatens our native species: the harlequin ladybird devours over 200

Spring clean to protect butterflies

Some butterfly species such as swallowtails hibernate as chrysalises stuck to a plant stem. This means an autumn clean in the garden not only removes dead stems but also numerous lodgers. It's better to cut back stems in spring and to store the cut stems in a dry, well-ventilated spot.

↑ Earwigs also appreciate decorative architecture.

aphids and mealy bugs per day. This conspicuous type of ladybird with its highly varying colouration profits less from a ladybird box: when it's time to hibernate, harlequin ladybirds gather in great numbers on exterior walls and then enter the building via cracks. This phenomenon has earned them the reputation as a plague, although they are harmless creatures.

Butterflies

Who doesn't love having butterflies in their garden? Unfortunately, they are an increasingly rare sight as their numbers are in serious decline. Butterflies spend the winters in various ways: some species migrate south, whereas others hibernate in sheltered spots either as an egg, caterpillar, chrysalis or adult butterfly. In the UK, six butterfly species hibernate in their adult state, and you can protect them from predators by providing suitable accommodation.

Butterflies have a particularly long proboscis to suck nectar from blossoms that remain inaccessible for bees, which is why 'butterfly flowers' such as bluebeard, purple rock cress (*Aubrieta*),

phlox, butterfly bush (*Buddleja*) and herb plants are particularly attractive. The caterpillars rely on specific forage plants: certain grasses, blackberry and raspberry, thistles, stinging nettles, papilionaceous flowers such as sweet pea and others, mustards and willows.

⬇ Potential inhabitants of butterfly boxes include peacock butterflies.

Earwig's paradise

Time ½ hour
Build very easy

Why only ever use standard flower pots? Our earwig hotels are highly decorative in addition to their usefulness.

1 It can't get any easier: loosely fill your 15–20cm high favourite container with wood shavings or straw.

2 If you're using a container with a very smooth surface that widens towards the opening, you should affix wire mesh across the opening to avoid the filling sliding out. You can simply bend the wire mesh over the edge and squeeze it tight or secure it with sisal garden twine.

3 Hang up your earwig hotels with the opening facing down into fruit trees, on your fence or on sticks in your beds.

The pots should not hang from long pieces of twine as the entrance at the bottom has to be easy to reach for these insects and they are not the best of climbers.

> **Materials**
> * small terracotta plant pots or zinc buckets
> * wood shavings or straw
> * galvanised wire mesh

Green lacewing box

Time 3 hours
Build easy

Green lacewings love red, so boxes for these delicate green critters should be bright red.

1 You have two options for designing the entrance: the easiest is to cut out an oblong or round hole and affix galvanised wire mesh across it using a stapler. Or you could saw several 2cm high horizontal slits into the front panel.

2 Glue the parts together as shown in the illustration and, after pre-drilling the screw holes, screw it all together.

3 Attach the door with two rustproof hinges and affix a hook and screw eye for closing it.

4 Paint the exterior of the box with environmentally friendly red paint.

5 Loosely fill the box with wheat straw and hang it up in a dry, sheltered spot.

Plant catmint and late-summer flowers such as yarrow (*Achillea*) and purple coneflower (*Echinacea purpurea*) in your garden: green lacewings love these!

Materials
- bright red, environmentally friendly weatherproof paint
- galvanised wire mesh, stapler, if needed
- rust-resistant screws 30mm
- 2 rust-resistant hinges and screws
- rust-resistant hook and screw eye to close the lid
- wheat straw
- jigsaw, drill, cordless screwdriver

Quantity	Name	Material	Height/ length (mm)	Width (mm)
1	floor	plywood 15mm	180	200
1	front door	see above	250	200
2	side panels	see above	235	180
1	back panel	see above	250	200
1	roof	see above	220	220

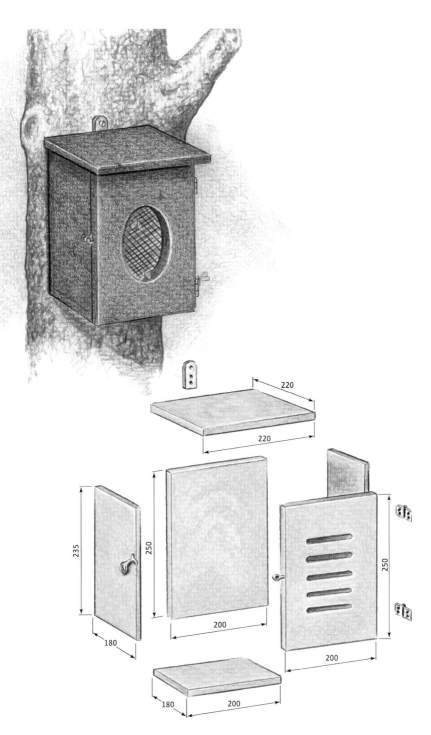

220
220
235
250
180
250
200
200
180
200

Ladybird hotel

Time 2 hours
Build easy

*Dry leaves as a filling
are particularly natural
– try it!*

1 The entrance holes are located on the box floor near the back, making it easy for ladybirds to crawl inside from the tree trunk or exterior wall. Drill several holes of about 1cm in diameter. If necessary, pre-drill and expand the hole with a coping saw or round file.

2 Cut the side panels to make a 75° angle (see illustration). Cut the upper edges of the front and back panel likewise in a 75° angle using a jigsaw.

3 Glue together the floor, side panels, front and back panel and secure everything with thin, rust-resistant screws after pre-drilling the screw holes.

4 Attach the lid with the hinges on the front panel so you can open it.

5 Loosely fill the box with dry leaves (beech or oak) or with wood shavings or straw.

Hang up the hotel on the trunk of a fruit tree or on a sunny exterior wall.

Materials

- 2 rust-resistant hinges, max. width 15mm, with screws
- straw, wood shavings or dry leaves
- rust-resistant screws 30mm
- jigsaw, coping saw, drill, cordless screwdriver

Quantity	Name	Material	Height/ length (mm)	Width (mm)
1	floor	plywood 15mm	200	300
2	side panels	see above	200	170
1	front panel	see above	155	300
1	back panel	see above	200	300
1	roof	see above	240	380

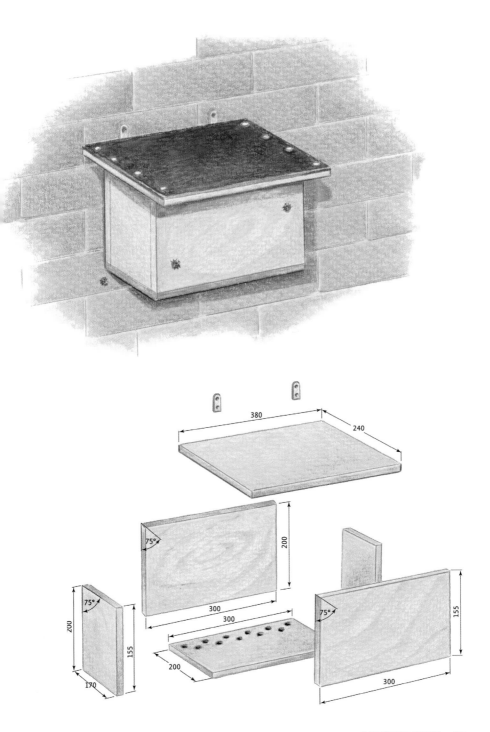

Butterfly shelter

Time 2–3 hours
Build easy

The butterfly shelter features a removable roof. However, as the butterflies often spend the winter hanging from the roof, take it off only in the warmer months.

1 First saw the front and back panel as shown in the illustration – the roof forms a right angle.
2 Saw slits of 12mm in width and 80mm in height into the front panel; it's best to pre-drill and then continue with a jigsaw. Smooth the edges with sandpaper.
3 Glue together the floor and walls and secure it all with rust-resistant screws after pre-drilling the screw holes.
4 Glue and screw together the two roof panels to form a right angle. Glue the three battens on to the roof as shown in the illustration so that the roof fits firmly and safely on the house.
5 Now place 2–3 thin branches inside the house to provide a perch for the butterflies. Make sure you leave enough free space for the insects.

Hang up your butterfly shelter in a medium shady spot away from the wind and weather.

Materials
- 2–3 branches
- rust-resistant screws 30mm
- jigsaw, drill, cordless screwdriver

Quantity	Name	Material	Height/ length (mm)	Width (mm)
1	floor	plywood 15mm	200	180
1	front panel	see above	350	200
2	side panels	see above	250	180
1	back panel	see above	350	200
1	roof panel	see above	230	150
1	roof panel	see above	230	135
3	battens for roof	wooden battens 1.5 x 1.5cm	180	

Index

Haynes Bee Manual

The *Bee Manual* provides a complete and easy-to-follow reference to the intriguing world of the honey bee and the addictive craft of beekeeping. Aimed at the novice but also containing plenty to interest the experienced beekeeper, the *Bee Manual* presents no-nonsense advice, facts and step-by-step sequences, as well as plenty of relevant photographs and diagrams. Find out how to work with these fascinating insects to enable them to thrive, carry out their pollination activities and produce a satisfying honey crop – and you could also play a part in reversing the decline in the number of bee colonies.

Haynes

Bee
Manual
The complete step-by-step guide to keeping bees

Claire & Adrian Waring
Foreword by Bill Turnbull

- The honey bee, its life cycle and activities, and how the colony operates.
- The honey bee in the environment and its importance in pollination.
- Choosing equipment and obtaining bees.
- Finding and laying out an apiary.

- How to inspect a colony, handle frames and control the bees.
- Advice on swarm prevention and control.
- Dealing with the honey crop.
- Uses for honey, beeswax and propolis.
- Detection and control of the main honey bee pests and diseases.

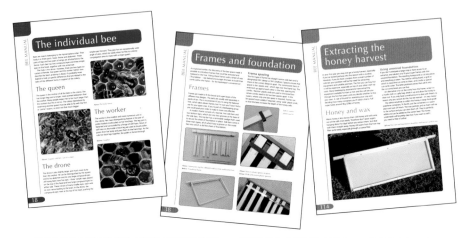